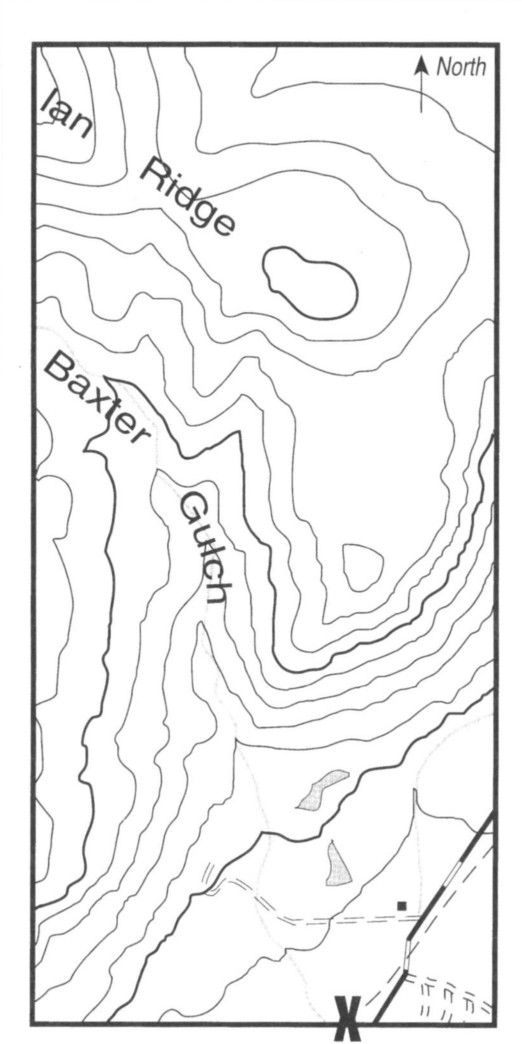

Standing at the "X" on the topo map (above left), someone looking north would see the scene depicted above, right, including the secondary highway, streams, house, unfinished roads, ponds, and mountain ridges.

Figure 1 • Topo Map View vs. Actual View

The Problem of Visualization

Many people have a difficult time visualizing three dimensions from two. In fact, some have a terrible time transferring from any drawing to the real thing. A number of your students may never, in the time you have available, develop the ability to look at a flat map and see a three-dimensional surface.

This problem is rather surprising because we live in a world filled with two-dimensional images of three-dimensional objects. Every day, most people view and understand photographs and television images as representations of three dimensions, yet maps still present a problem. This ease of transition may be because the objects on TV or in a photograph are familiar and those on a map are not, but the problem remains even when the map being used is of the student's home town area. A more likely possibility is that they are confused by the point of view.

Maps look straight down on Earth's surface and people rarely look at Earth that way even from an airplane. Our views are nearly always oblique—that is, from angles which are not parallel or perpendicular. (See Figure 1, above.) Show students an oblique photo of a landform near their homes and very few have difficulty identifying the location.

Topographic Mapping Symbols

Primary highway, hard surface
Secondary highway, hard surface
Light-duty road, hard or improved surface
Unimproved road
Trail
Railroad: single track
Railroad: multiple track
Bridge
Drawbridge
Tunnel
Footbridge
Overpass/Underpass
Power transmission line with located tower
TELEPHONE — Landmark line (labeled as to type)

Boundary: national
Boundary: state
Boundary: county, parish, municipio
Boundary: civil township, precinct, town, barrio
Boundary: incorporated city, village, town, hamlet
Boundary: reservation, national or state
Boundary: small park, cemetery, airport, etc.
Boundary: land grant
Township or range line, U.S. land survey
Section line, U.S. land survey
Township line, not U.S. land survey
Section line, not U.S. land survey
Fence line or field line
Section corner: found/indicated
Boundary monument: land grant/other

Dam with lock
Canal with lock
Large dam
Small dam: masonry/earth
Buildings (dwelling, workplace, etc.)
Cem — School/Church/Cemeteries
Buildings (barn, warehouse, etc.)
Water Tank — Tanks; oil, water, etc. (labeled only if water)
Oil, Gas — Wells other than water (labeled as to type)
U.S. mineral or location monument/Prospect
Quarry/Gravel pit
Mine shaft/Tunnel or cave entrance
Campsite/Picnic area
Located or landmark subject/Windmill
Exposed wreck
Rock or coral reef
Rock: bare or awash

Index contour
Supplementary contour
Mine dump
Dune area
Sand area
Tailings

Intermediate contour
Depression contours
Levee
Large wash
Tailings pond
Distorted surface
Gravel beach

Glacier
Perennial streams
Water well/Spring
Rapids
Channel
20 — Sounding/Depth curve
Dry lake bed

Intermittent stream
Aqueduct tunnel
Falls
Intermittent lake
Small wash
Marsh (swamp)
Land subject to controlled inundation

Horizontal control station
BM ×545, ×546 — Vertical control station
213, +76 — Road fork/Section corner with elevation
× 2650 — Checked spot elevation
× 2650 — Unchecked spot elevation

Woodland
Submerged marsh
Orchard
Vineyard

Mangrove
Scrub
Wooded marsh
Bldg. omission area

From *Topographic Map Symbols* (Issued by the U.S. Department of the Interior/U.S. Geological Survey) Item No. TUS0669

Topographic Map Symbols

What is a Topographic Map?

A map is a representation of the Earth, or part of it. The distinctive characteristic of a topographic map is that the shape of the Earth's surface is shown by contour lines. Contours are imaginary lines that join points of equal elevation on the surface of the land above or below a reference surface, such as mean sea level. Contours make it possible to measure the height of mountains, depths of the ocean bottom, and steepness of slopes.

A topographic map shows more than contours. The map includes symbols that represent such features as streets, buildings, streams, and vegetation. These symbols are constantly refined to better relate to the features they represent, improve the appearance or readability of the map, or reduce production cost.

Consequently, within the same series, maps may have slightly different symbols for the same feature. Examples of symbols that have changed include built-up areas, roads, intermittent drainage, and some lettering styles. On one type of large-scale topographic map, called provisional, some symbols and lettering are hand-drawn.

Reading Topographic Maps

Interpreting the colored lines, areas, and other symbols is the first step in using topographic maps. Features are shown as points, lines, or areas, depending on their size and extent. For example, individual houses may be shown as small black squares. For larger buildings, the actual shapes are mapped. In densely built-up areas, most individual buildings are omitted and an area tint is shown. On some maps, post offices, churches, city halls, and other landmark buildings are shown within the tinted area.

The first features usually noticed on a topographic map are the area features, such as vegetation (green), water (blue), and densely built-up areas (gray or red).

Many features are shown by lines that may be straight, curved, solid, dashed, dotted, or in any combination. The colors of the lines usually indicate similar classes of information: topographic contours (brown); lakes, streams, irrigation ditches, and other hydrographic features (blue); land grids and important roads (red); and other roads and trails, railroads, boundaries, and other cultural features (black). At one time, purple was used as a revision color to show all feature changes. Currently, purple is not used in our revision program, but purple features are still present on many existing maps.

Various point symbols are used to depict features such as buildings, campgrounds, springs, water tanks, mines, survey control points, and wells. Names of places and features are shown in a color corresponding to the type of feature. Many features are identified by labels, such as "Substation" or "Golf Course."

Topographic contours are shown in brown by lines of different widths. Each contour is a line of equal elevation; therefore, contours never cross. They show the general shape of the terrain. To help the user determine elevations, index contours are wider. Elevation values are printed in several places along these lines. The narrower intermediate and supplementary contours found between the index contours help to show more details of the land surface shape. Contours that are very close together represent steep slopes. Widely spaced contours or an absence of contours means that the ground slope is relatively level. The elevation difference between adjacent contour lines, called the contour interval, is selected to best show the general shape of the terrain. A map of a relatively flat area may have a contour interval of 10 feet or less. Maps in mountainous areas may have contour intervals of 100 feet or more. The contour interval is printed in the margin of each U.S. Geological Survey (USGS) map.

Bathymetric contours are shown in blue or black, depending on their location. They show the shape and slope of the ocean bottom surface. The bathymetric contour interval may vary on each map and is explained in the map margin.

U.S. Department of the Interior
U.S. Geological Survey

BATHYMETRIC FEATURES

Area exposed at mean low tide; sounding datum line***

Channel***

Sunken rock***

BOUNDARIES

National

State or territorial

County or equivalent

Civil township or equivalent

Incorporated city or equivalent

Federally administered park, reservation, or monument (external)

Federally administered park, reservation, or monument (internal)

State forest, park, reservation, or monument and large county park

Forest Service administrative area*

Forest Service ranger district*

National Forest System land status, Forest Service lands*

National Forest System land status, non-Forest Service lands*

Small park (county or city)

BUILDINGS AND RELATED FEATURES

Building

School; house of worship

Athletic field

Built-up area

Forest headquarters*

Ranger district office*

Guard station or work center*

Racetrack or raceway

Airport, paved landing strip, runway, taxiway, or apron

Unpaved landing strip

Well (other than water), windmill or wind generator

Tanks

Covered reservoir

Gaging station

Located or landmark object (feature as labeled)

Boat ramp or boat access*

Roadside park or rest area

Picnic area

Campground

Winter recreation area*

Cemetery — Cem

COASTAL FEATURES

Foreshore flat — Mud

Coral or rock reef — Reef

Rock, bare or awash; dangerous to navigation

Group of rocks, bare or awash

Exposed wreck

Depth curve; sounding — 18 23

Breakwater, pier, jetty, or wharf

Seawall

Oil or gas well; platform

CONTOURS

Topographic

Index — 6000

Approximate or indefinite

Intermediate

Approximate or indefinite

Supplementary

Depression

Cut

Fill

Continental divide

Bathymetric

Index***

Intermediate***

Index primary***

Primary***

Supplementary***

CONTROL DATA AND MONUMENTS

Principal point** — 3-20

U.S. mineral or location monument — USMM 438

River mileage marker — Mile 69

Boundary monument

Third-order or better elevation, with tablet — BM 9134 — BM 277

Third-order or better elevation, recoverable mark, no tablet — 5628

With number and elevation — 67 4567

Horizontal control

Third-order or better, permanent mark — Neace — Neace

With third-order or better elevation — BM 52 — Pike BM393

With checked spot elevation — 1012

Coincident with found section corner — Cactus — Cactus

Unmonumented**

CONTROL DATA AND MONUMENTS – *continued*

Vertical control

Third-order or better elevation, with tablet	BM × 5280
Third-order or better elevation, recoverable mark, no tablet	× 528
Bench mark coincident with found section corner	BM + 5280
Spot elevation	× *7523*

GLACIERS AND PERMANENT SNOWFIELDS

- Contours and limits
- Formlines
- Glacial advance
- Glacial retreat

LAND SURVEYS

Public land survey system

Range or Township line	
Location approximate	
Location doubtful	
Protracted	
Protracted (AK 1:63,360-scale)	
Range or Township labels	R1E T2N R3W T4S
Section line	
Location approximate	
Location doubtful	
Protracted	
Protracted (AK 1:63,360-scale)	
Section numbers	1 - 36 1 - 36
Found section corner	
Found closing corner	
Witness corner	WC
Meander corner	MC
Weak corner*	

Other land surveys

- Range or Township line
- Section line
- Land grant, mining claim, donation land claim, or tract
- Land grant, homestead, mineral, or other special survey monument
- Fence or field lines

MARINE SHORELINES

- Shoreline
- Apparent (edge of vegetation)***
- Indefinite or unsurveyed

MINES AND CAVES

Quarry or open pit mine	
Gravel, sand, clay, or borrow pit	
Mine tunnel or cave entrance	
Mine shaft	
Prospect	X
Tailings	Tailings
Mine dump	
Former disposal site or mine	

PROJECTION AND GRIDS

Neatline	39°15′ 90°37′30″
Graticule tick	55′
Graticule intersection	
Datum shift tick	

State plane coordinate systems

Primary zone tick	640 000 FEET
Secondary zone tick	247 500 METERS
Tertiary zone tick	260 000 FEET
Quaternary zone tick	98 500 METERS
Quintary zone tick	320 000 FEET

Universal transverse metcator grid

UTM grid (full grid)	273
UTM grid ticks*	269

RAILROADS AND RELATED FEATURES

- Standard guage railroad, single track
- Standard guage railroad, multiple track
- Narrow guage railroad, single track
- Narrow guage railroad, multiple track
- Railroad siding
- Railroad in highway
- Railroad in road
- Railroad in light duty road*
- Railroad underpass; overpass
- Railroad bridge; drawbridge
- Railroad tunnel
- Railroad yard
- Railroad turntable; roundhouse

RIVERS, LAKES, AND CANALS

- Perennial stream
- Perennial river
- Intermittent stream
- Intermittent river
- Disappearing stream
- Falls, small
- Falls, large
- Rapids, small
- Rapids, large
- Masonry dam
- Dam with lock
- Dam carrying road

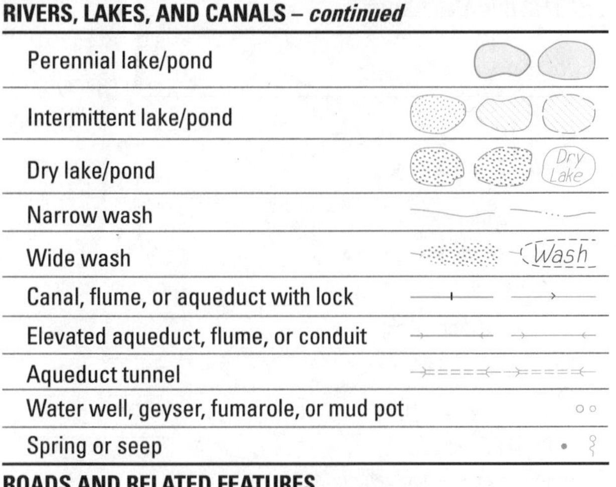

ROADS AND RELATED FEATURES

Please note: Roads on Provisional-edition maps are not classified as primary, secondary, or light duty. These roads are all classified as improved roads and are symbolized the same as light duty roads.

- Primary highway
- Secondary highway
- Light duty road
- Light duty road, paved*
- Light duty road, gravel*
- Light duty road, dirt*
- Light duty road, unspecified*
- Unimproved road
- Unimproved road*
- 4WD road
- 4WD road*
- Trail
- Highway or road with median strip
- Highway or road under construction — Under Const
- Highway or road underpass; overpass
- Highway or road bridge; drawbridge
- Highway or road tunnel
- Road block, berm, or barrier*
- Gate on road*
- Trailhead* — TH

SUBMERGED AREAS AND BOGS

- Marsh or swamp
- Submerged marsh or swamp
- Wooded marsh or swamp
- Submerged wooded marsh or swamp
- Land subject to inundation — Max Pool 431

SURFACE FEATURES

- Levee — Levee
- Sand or mud — Sand
- Disturbed surface
- Gravel beach or glacial moraine — Gravel
- Tailings pond — Tailings Pond

TRANSMISSION LINES AND PIPELINES

- Power transmission line; pole; tower
- Telephone line — Telephone
- Aboveground pipeline
- Underground pipeline — Pipeline

VEGETATION

- Woodland
- Shrubland
- Orchard
- Vineyard
- Mangrove — Mangrove

* USGS-USDA Forest Service Single-Edition Quadrangle maps only.
In August 1993, the U.S. Geological Survey and the U.S. Department of Agriculture's Forest Service signed an Interagency Agreement to begin a single-edition joint mapping program. This agreement established the coordination for producing and maintaining single-edition primary series topographic maps for quadrangles containing National Forest System lands. The joint mapping program eliminates duplication of effort by the agencies and results in a more frequent revision cycle for quadrangles containing National Forests. Maps are revised on the basis of jointly developed standards and contain normal features mapped by the USGS, as well as additional features required for efficient management of National Forest System lands. Single-edition maps look slightly different but meet the content, accuracy, and quality criteria of other USGS products.

** Provisional-Edition maps only.
Provisional-edition maps were established to expedite completion of the remaining large-scale topographic quadrangles of the conterminous United States. They contain essentially the same level of information as the standard series maps. This series can be easily recognized by the title "Provisional Edition" in the lower right-hand corner.

*** Topographic Bathymetric maps only.

Topographic Map Information

For more information about topographic maps produced by the USGS, please call:
1-888-ASK-USGS or visit us at http://ask.usgs.gov/

ISBN 0-607-96942-3

Printed on recycled paper

Show the same students an aerial photo of an equally familiar area and many will be clueless. It may help you to create a bulletin board display of the local area using maps plus aerial and oblique photos. (See the Resources section for information on how and where to obtain aerial photos.)

Why Teach Topo Map Skills?

With that caution in mind, is it worth the time and effort to teach topographic map skills? It is indeed, precisely because of the problems students encounter trying to visualize three dimensions from two. Once students begin to grasp this concept, they will be able to use this skill in other areas quite remote from Earth science, such as visualizing a complete three dimensional house from a set of two-dimensional house plans or a completed dress or shirt from a pattern.

More directly, an ability to read topo maps might help students: understand land use, highway, and railroad planning as well as flood plain and flooding problems; identify geographical features such as mountains and stream systems; and learn about routes of historical trails, forts, battles, and settlements, as well as political boundaries. Many ghost towns which have been dropped from typical highway maps are still listed on USGS topo maps.

In the Earth sciences, topo maps are especially valuable in dealing with landforms such as mountains, stream systems, plateaus, glacial features, and plains because of their ability to depict three-dimensional features. If students know the basics of landforms, working with topo maps will reinforce and allow them to apply their knowledge.

Figure 3 • A Reminder About Latitude and Longitude

A degree of latitude or longitude is actually a measurement of arc following the curve of Earth. Mapmakers devised latitude and longitude so they could accurately represent Earth on their maps. Lines of latitude—called parallels—circle the globe parallel to each other and are measured in degrees north or south of the equator. The equator is the 0° parallel, circling at the middle of the globe. Longitude lines—or meridians—are lines from pole to pole and are measured east and west from the Prime Meridian to the International Date Line.

One degree of latitude or longitude—or arc—contains 60 minutes of arc. Further, one minute of arc contains 60 seconds of arc. An example of the coordinates for a specific location somewhere on the Earth would be 45° 21' 34" N by 109° 38' 3"W. The first string of numbers is "45 degrees, 21 minutes, 34 seconds north latitude" and the second is "109 degrees, 38 minutes, 3 seconds west longitude."

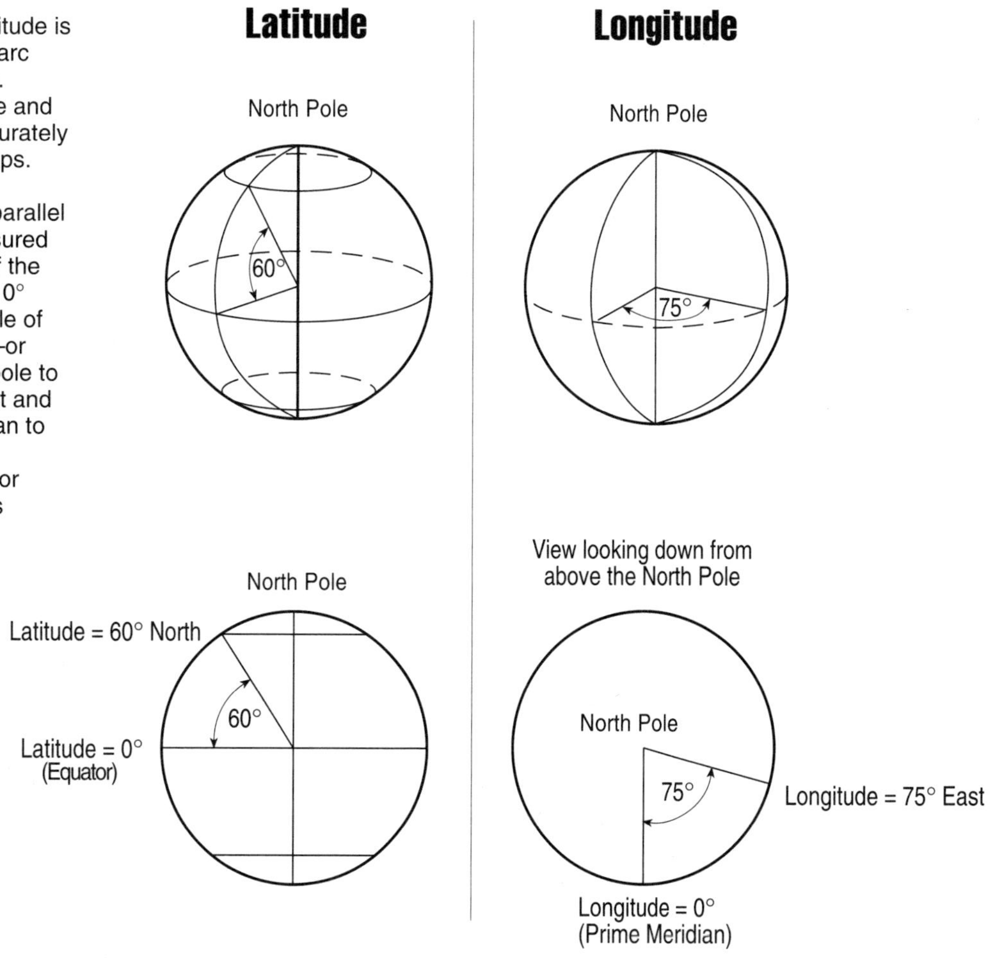

Touring a Topographic Map

If you have not spent a great amount of time with topographic maps, this section may be a valuable review. If you are already quite familiar with them, skip on to the section titled "The First Step."

Map reading always starts with the information found in the margins. (See Figure 8 on page 12 for a detailed diagram of the information in the margins of a topo map.) On a topo map, the top margin lists who produced the map, and, in the upper right, the name of the map and the series. The map name is usually the same as a prominent feature—a town, city, or physical feature such as a mountain peak—in the map area. The map accompanying this publication is likely to be a 7.5- or 15-minute series. (These numbers refer to the dimensions of the topo map. A 7.5-minute series map is 7.5 minutes of latitude by 7.5 minutes of longitude. See Figure 3: A Reminder About Latitude and Longitude on page 5.)

To check the series of your topo map, subtract the latitude of the lower right corner from the latitude of the upper right corner. This number should the same as the series number. Likewise, subtracting the smaller longitude from the larger should produce the same number.

Because these maps have four equal sides in terms of degrees of latitude and longitude, they are called *quadrangles* or *quads*. At this point one is faced with the realization that a degree of latitude is not the same length as a degree of longitude. The length of a degree of latitude stays the same worldwide but degrees of longitude differ depending upon distance from the equator. Confused? Look at a globe. At the north and south poles, a degree of longitude becomes infinitely small.

The bottom margin holds a wealth of information. One important piece of data is the declination—the difference between true north and magnetic north. A compass in most of North America will not point to true north—true north being the north pole. The indicator in the bottom margin shows the difference between true north and magnetic north for people using the map for precise location. Further explanation of this difference is not necessary for most classroom uses of the maps.

A Quick Review of Scale

Another important element in the bottom margin is the scale of the map. The scale of the map which accompanies this booklet is likely to be 1:24,000—that is, a 7.5-minute series map. This means that one unit of measurement on the map is equal to 24,000 of the same units on the Earth's surface. For example, 1 cm on the map equals 24,000 cm on Earth. It may be easier to think of the scale as saying that the map is 1/24,000th the size of the area it represents. The larger the second number in the ratio—in this case 24,000—the less detail the map can show. Scales are also used by architects and model makers when they create scaled-down versions of houses, cars, or airplanes.

The entire country has been mapped at a scale of 1:250,000. These maps show an area of 1° latitude by 2° longitude. They are useful because each one shows a much larger portion of the countryside and contain amazing detail considering the scale. They are fine for general work but are difficult for students to use for detailed investigation. One advantage of the 1:250,000 maps is that matching plastic raised relief maps are available. You might want to consider using these maps to aid students in the visualization step. Originally produced by the Army Map Service, the raised relief maps are now available commercially.

The bottom margin also includes sample linear scales in various units, the contour interval (more about this later), the name and publication date of the map, and technical information about the production of the map.

Inspecting the Map

Now let's look at the actual map. If your map is of an area west of Ohio, it is likely that there are vertical and horizontal lines crossing it. To interpret these, look in the margin at either end of the line. The numbers, or numbers and letters, tell what the line indicates. If the numbers are degrees, minutes, and seconds, or just minutes and seconds, the line must be one of either latitude (horizontal) or longitude (vertical). If the label is something like T33N or R78W then the line indicates a change of township or range.

Township and *range* are terms used in a surveying system throughout much of the United States west of Ohio—the government land office grid system. As territories were claimed by and later joined the United States, they were divided into units of land measuring six miles on a side. These units were numbered north or south of a selected parallel of latitude called a *base line* and east or west of a selected meridian of longitude called a *principal meridian*. In each territory, north or south measurement is called *township* and east or west measurement is called *range*. The actual six-mile by six-mile area of land is called ***a** township*. (See Figure 4, below.) Each township was sub-divided into 36 1-mile by 1-mile units called sections. Sections are identified by a number—starting with 1 in the NE corner of the township. The diagram shows the numbering pattern. Look at the map and see if there is a grid of squares about 3 inches on a side which, generally, have a number in the center. (Remember, if your map is of an area east of Ohio it will probably not have these markings.) If so, these are sections. They are handy for estimating distance or area. The land

Figure 4 • Township and Range Grid Pattern

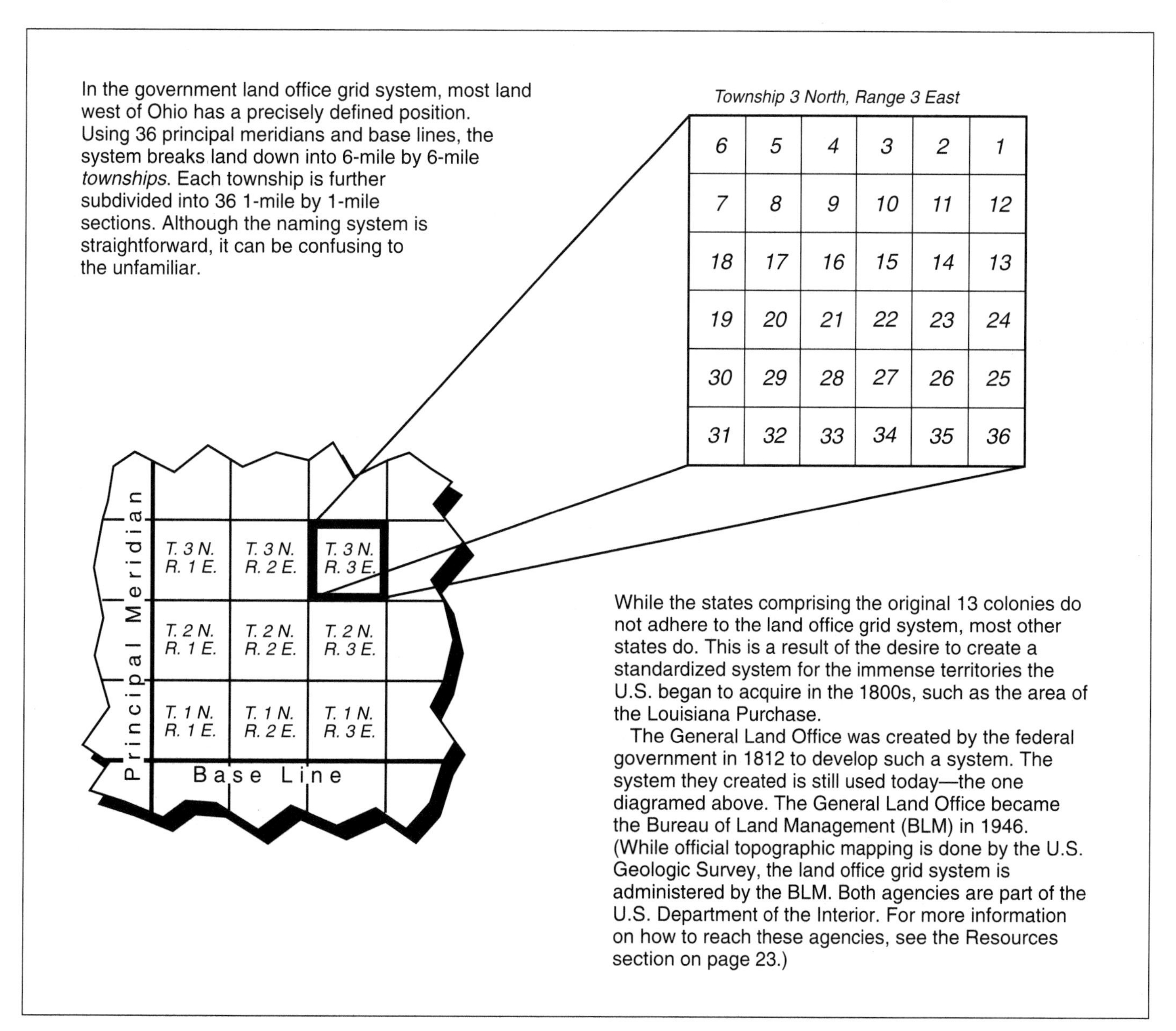

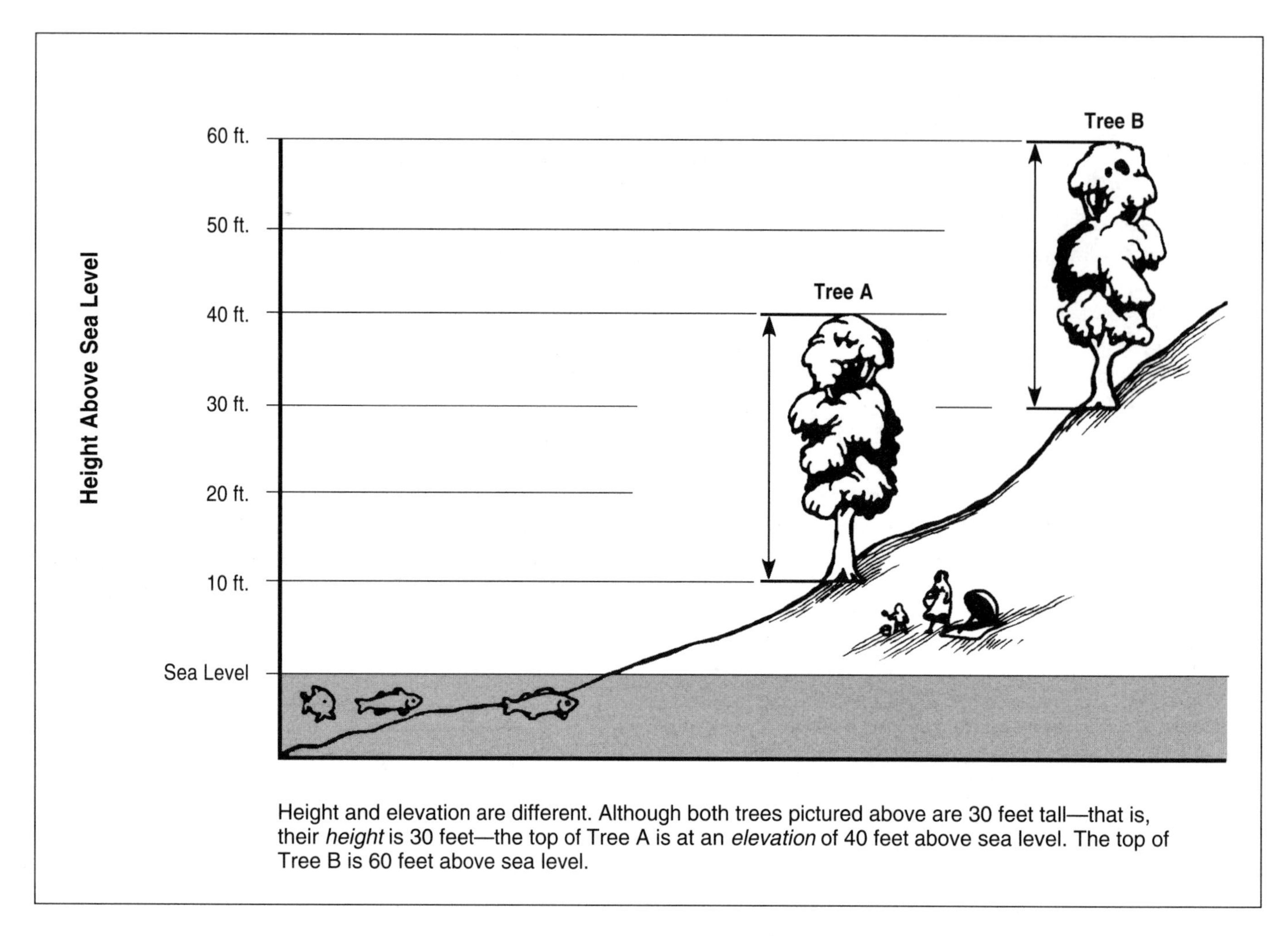

Height and elevation are different. Although both trees pictured above are 30 feet tall—that is, their *height* is 30 feet—the top of Tree A is at an *elevation* of 40 feet above sea level. The top of Tree B is 60 feet above sea level.

Figure 5 • Height vs. Elevation

office grid system is important as it is used for legal land description.

The map itself contains the usual map features—towns, roads, etc.—in various colors and sizes to indicate importance. One feature, however, may not be familiar to students. It sets topographic maps apart from all other maps and it is why you are reading this booklet. The special feature is squiggly brown lines running all over the map. These lines are called *contour lines*.

Contour Lines

Look at the contour lines on your map. Note that some of them are thicker than others. Select one of these thicker lines and follow it. Somewhere along the line there will be a number, also in brown, written in the direction of and in place of the line. This is the elevation of the line. Elevation is the height—in feet or meters—above sea level of a particular point or line. (See Figure 5, above.) Every point along that line is at the same elevation—as shown by the brown number. In the past, a surveyor in the field would measure and plot a number of points at the same elevation and then connect these points with a line. This led to the definition of a contour line as being a line connecting points of equal elevation. Today, most of the work is done with aerial photos and the definition, though accurate, has less significance. Two points to remember: a contour line *never* goes up or down hill and contour lines *never* cross each other.

The vertical distance between contour lines is called the *contour interval*. (Remember it from the bottom margin?) Going from one contour line to the next, one goes up or down one contour interval. The simplest way to decide if it is up or down is to check labels on the thick contour lines. The contour interval used varies from map to map, depending upon the terrain. In Wyoming, most topo maps have a contour interval of 20 feet, in Florida a contour interval of 2 feet is appropriate, and a topo map for an area in the Rocky Mountains west of Denver lists 80 feet as the contour interval.

By locating the closest labeled contour line and then counting lines, one can determine the elevation of a point. However, resist the temptation to guess between lines. (See Figure 6, below.) The map tells exactly what the elevation is at a specific line, but it does not tell what happens to the land area between the lines. The only safe way to give the elevation of a point between the lines is to give the elevation of both adjoining lines. For example, between 4,020 feet and 4,040 feet.

The most important rule to know about topo maps is also the easiest to remember: "The closer together the contour lines, the steeper the slope." Using this rule one can take a glance at a topographic map and have some idea of the terrain. It becomes more significant if one looks at the contour interval first. A few contour lines close together with a contour interval of 2 feet do not indicate the same sort of obstacle they would if the contour interval was 80 feet.

One other handy bit of information: when contour lines cross a stream they form a "V" that points upstream. (This is known as the rule of the "V"s and is investigated further in Activity Ten: Down the Drainage Plan.) Knowing that, it is easy to tell the direction of stream flow. Ask students why this is true and see if they can explain.

The First Step

It is probably worth your time to construct a three-dimensional model from a topographic map so students have a middle step in their visualization. Using the typical plastic raised relief map helps but they are too smooth to make the idea of the lines stand out. A built-up map can be constructed without much difficulty by using foam core board (1/4"-thick board makes the vertical and horizontal scales about the same when working on a 1:24,000 scale map.) Foam core makes a better looking map than corrugated cardboard. With either medium, lots of time will be spent in cutting and gluing and one might as well invest a few dollars to have a nice looking product. Instead of trying to make an entire map, it may make more sense to build up a quarter or so of the map that is the most interesting. A small segment will get the point across as well as a complete map.

Figure 6 • The Problem with Predicting

The figure at right shows why the only safe way of giving the elevation of a point not directly on a contour line is by saying, the point lies between the elevations of the two nearest contour lines. The dashed line represents a predicted elevation. But as the actual profile—the solid line—shows, the terrain can vary from the predicted line. In the bottom example, the actual elevation of the midway point between 4000 feet and 4020 feet is about 4003 feet, not the halfway value of 4010 feet that the predicted line shows. The upper example shows the same possible error, this time the predicted elevation—4030 feet—is eight feet less than the actual elevation at the midway point, which is 4038 feet. Topo maps offer an enormous amount of information, but readers can't "guess" between the lines.

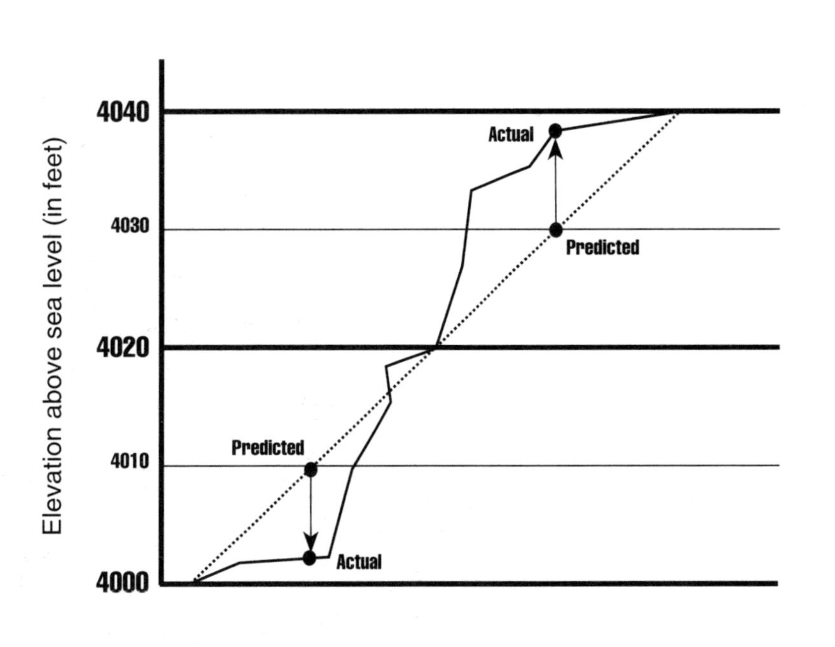

Figure 7 • Assembling a 3D Topo Map

Directions for a 3D model of a topo map

Step 1: To make the foam core map, select a relatively easy portion of the map. For example, do not pick an area where the contour lines are so close together that tracing is difficult.

Step 2: Either make photocopies or place a piece of tracing paper over the map and trace only the thicker contour lines—the lines with elevation numbers in them.

Step 3: Use carbon paper to transfer each dark contour line to a different piece of foam core. Also mark the next layer with a dashed line. (This will help you align the foam core pieces later on.) It is a good idea to label elevations on each piece to help when stacking the layers. Don't forget a base piece which represents the lowest elevation in the area.

Step 4: Cut along the solid line on each piece, beginning with the largest piece first. (Remember the dashed line is for help in placement—not for cutting.) Cutting can be done with a knife but is much easier with a scroll saw. A very fine blade makes it possible to cut more accurately.

Step 5: Stack as you cut, making sure that the pieces are aligned.

Step 6: Glue the pieces together, working from the smallest, top pieces down. White glue works fine.

Once you have a sample for students to see, some of them may want to do built-up maps of their own for the classroom. A set of these showing the vicinity of your school is a valuable asset. If you don't have a topo map of your school area, maybe groups of students will want to work on different sections of the topo map you do have—adding them all together to create one large representation of the entire map.

Activities for Your Students

Once you feel familiar enough with the material in the introductory section above, the following activities will help your students explore and understand topo maps. It is highly recommended that there be a map for every three or four students in your class. (Maps are available from the USGS—see the Resources section at the end of this booklet for the address—and often from local map or camping stores.) Laminating the maps before you pass them out is a good idea. This will allow students to write on the maps with washable markers—the type used for transparencies.

If you can't get that many, you could photocopy the map that came with this booklet in sections, taping together the pieces to form a complete map. But be careful if you do this because the quality of the maps, especially the contour lines, may be so degraded that the maps are essentially unusable. Also, all colors become black on standard photocopies.

Each activity has two sections: the first is information about the level of the activity and the skills it will help develop as well as materials you will need. The second section is the actual activity. It directs you, the instructor, in approaches you might use and suggested questions you might ask. Don't feel confined by the language of the activity. If you want to ask for more information, or revisit past concepts, please do. The activities are to be used as guides, not absolutes.

Activity One

What's in a Name

This activity will help your students get acquainted with the map by asking them to locate some basic margin information. They will need to look at a map for the answers, so a map should be available for them to investigate. Be aware that your students probably will ask about other information in the margin. Be prepared to answer briefly and save a thorough discussion of it until the later, appropriate activity.

Ask your students to explore the margins for information. Direct their attention to the upper right of the map—where the map's title is located—and the lower right—which lists the date it was prepared. Keep in mind that if a map was prepared some years ago there may have been changes in the landscape since that time. The title of the USGS topographic map is always that of a prominent feature, often a town, located on the map. Challenge the class to find this title geographic feature on your map. Ask your students what they would guess the origin of the name is. Does it appear that this feature was named for a person or for some characteristic of the feature? In what state is the area depicted on the map located? Is anyone in your class from that state? The class might like to collect material about their map state—photos, videos, books, perhaps even rocks and plant material (or information about them), as well as other maps such as state highway maps, historical maps, or maps from an atlas.

What other information is there in the margin? See Figure 8 on page 12 for a diagram of what information exists where. Symbols, scale, and contour intervals will be treated in later activities.

Ask the class what direction north is on the map. Ask them how they know. Did they use the indicator on the map? Discuss the standard map representation of north being the top, south the bottom, east to the right, and west to the left. Is this true of other maps in the classroom? At home? Ask the students why almost all maps use this designation. Reinforce that many types of maps are used by many different people and a standard way of representing north is necessary. (Historically, the standard apparently developed in Europe in the 1500s as more and more maps were drawn and a standard was necessary.) Ask students to imagine what it would be like to try and read different maps today, if north was not standardized to be toward the top of a map. What problems might arise?

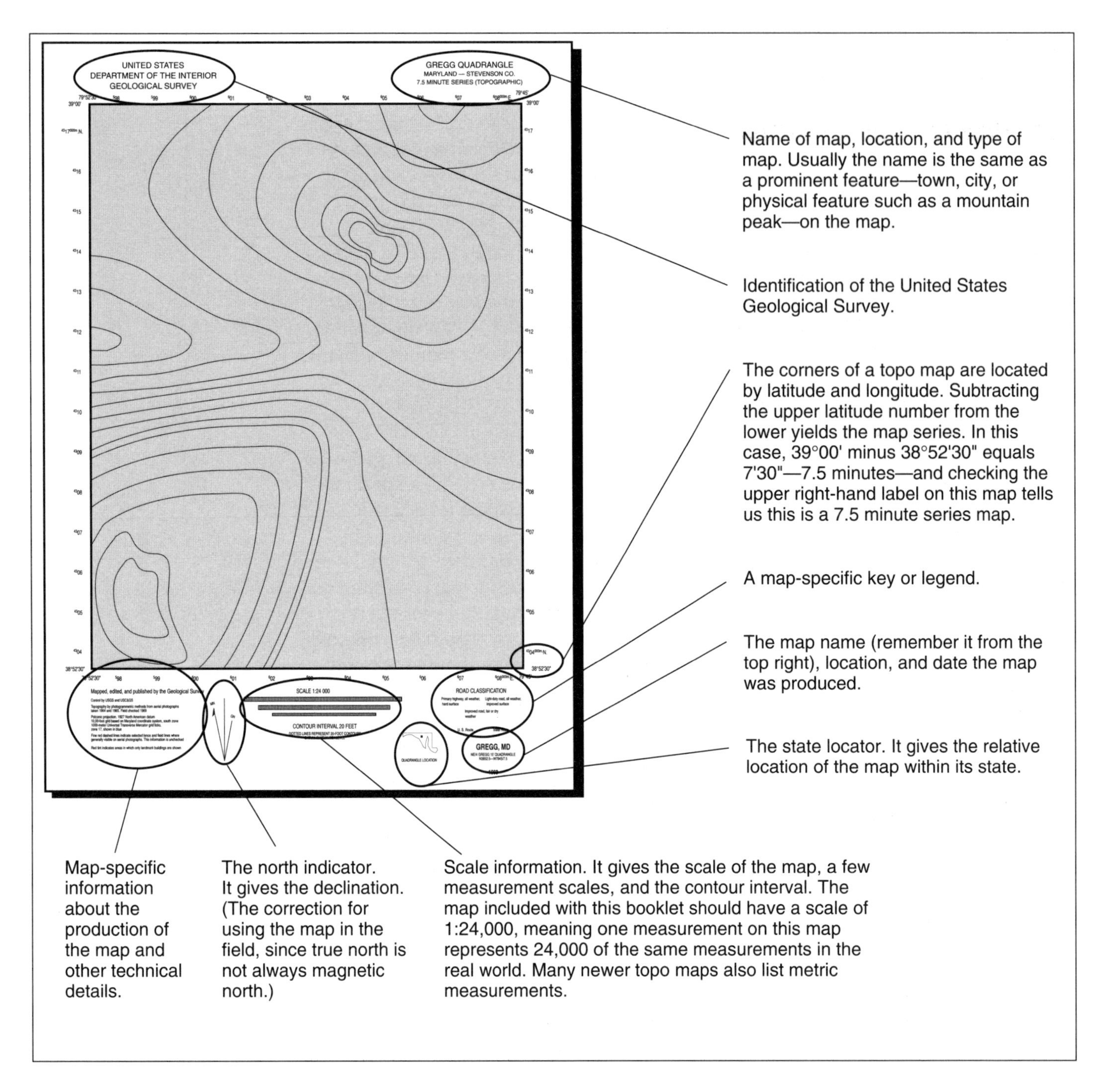

Figure 8 • Dissecting a Topographic Map

Activity Two

What Do All These Symbols Mean?

This activity introduces students to the symbols used on topo maps. (Nearly 100 symbols are defined by the USGS for use on topographic maps.) Students will need maps and copies of keys for symbols. (The key used in this booklet on page 4 lists nearly all the symbols students will encounter. The USGS also has keys.) Note that many of the symbols are in color. Since photocopiers do not reproduce color, some of the symbols may be confusing if the key is copied. Using original topo maps and sharing the key on page 4 will give the best results.

Topographic maps show many features. Ask the students to find an easy symbol, like a building. Then ask students to find as many symbols as they can on the map. Make a game out of it. Which team of four or five students can locate the most symbols? Which team can locate the most, different symbols? Create a master list of the symbols the students find. They may find symbols not identified by the key included in this booklet. If you don't have the full key from the USGS, ask your students what they think the symbols represent.

Students will likely query you about the brown contour lines and numbers. If they don't, direct their attention to them. Discuss elevation—distance (on these maps it's usually in feet) above sea level. Find the two ways elevations are marked on topo maps. The first is as a number, also in brown, and written in the direction of and in place of a contour line. This signifies the elevation of that contour line. The second is called a summit elevation. It is usually marked in brown next to a small brown "x." Summit elevations give an actual elevation reading of the spot where the "x" is. Usually these are mountain peaks or summits. What is the highest elevation anyone can find written in the numbers on the map? What is the lowest elevation? (Soon we will learn to interpret the contour lines.)

Activity Three

Understanding Scale

This activity explores the concept of scale. Again, students will need a topo map and measuring tools (rulers, meter sticks, tape measures).

If your students found the scale in the first activity, redirect their attention there now. Otherwise, draw your students' attention to the bottom margin. Tell students most maps have a scale. Ask them why? Ask them how useful a map without a scale would be? Would there be times when no scale is needed? Ask students to find the scale of the topo map. It's at the bottom center of the map. On the topo map that comes with this booklet, the scale will probably be 1:24,000. Explain that the scale is actually a *ratio*.

Define ratio as a relationship between two things. In this case, between the map and the real-life area the map represents. This means one of any unit, say one centimeter, on the map represents 24,000 of that unit on the ground or 24,000 cm. Mark 1 cm on the map or have the students mark it on their maps. Describe going to the actual site of the map and measuring 24,000 centimeters on the ground. Now encourage the students to see one finger length on the map as 24,000 finger lengths at the actual location. Have them invent things to use as measures—one hand length, one stick of gum, one shoelace, one black licorice whip, one small snake (imaginations, please) equals 24,000 of these actually laid end to end on the ground anywhere in your map area. Other topo map series will have different scales, but scale will always be the same within a series. A 15-minute series quadrangle has a scale of 1:62,500.

You may want to review the discussion of scale in the introduction for examples and wording.

Activity Four

Using Latitude and Longitude to Locate the Map

This exercise should be used only if students have already been introduced to longitude and latitude. Students will again need their topo maps, but this time they will also need additional maps that show a larger portion of Earth—like a state or a region. These larger maps should have latitude and longitude marked on them at intervals no greater than two degrees. If the intervals are larger it becomes difficult to locate a topo map with any accuracy.

Ask students how they would visit the site of the topo map. The students could use the name of any prominent feature (town, mountain) and see if it is on a larger atlas or state map. They might also come up with other ways of locating the site of the topo map. Direct the discussion to the best and most accurate way: using latitude and longitude. Locate a state map, perhaps from an atlas, that shows latitude and longitude. An encyclopedia map most likely will not do—detail will not be great enough. Ask the students to find the latitude and longitude of the topo map. The notations are at each corner of the map. Help students locate their topo map between two lines of latitude (top and bottom of the map) and two lines of longitude (left and right sides of the map). Most topographic maps list degrees, minutes, and seconds, thereby giving an exact location of the area they represent. Thus 45°10'11" means 45 degrees, 10 minutes, 11 seconds. It is important to help students have a sense of adventure in trying to locate the general area of the topographic map, rather than worrying about precision.

Activity Five

Where, Oh Where Does My Topo Map Belong? (Part 1)

This activity nicely follows the above activity, although they can be used independently. Students develop ways to travel to their map site(s) and study other aspects of the map site. Students will need their maps to locate the site, but otherwise much of the work is researching and investigating—almanacs, newspapers, TV news, and other materials generally available at the library.

Challenge students to plan a hypothetical class visit to your map state. Using other maps, find routes such as highways, rivers, railroads, coastlines, etc., students could use to get to the map area. Using the scale on a U.S. map and a ruler, ask students to determine how many miles away the map's state is from your own city and state. Ask students approximately how many hours it would take them to get there by different modes of travel. How fast will they travel? Maybe they're traveling by plane, bike, horseback, walking, or maybe supersonic jet. Divide the speed of travel into the number of miles. This may be several day's journey. This activity gives students an idea of rate, speed, and distance.

What time of year are the students traveling? What weather might they expect? Would it be similar to their own local weather? Have students check the weather section of the newspaper for listings of temperatures of cities across the country. Assign students to watch news programs on TV to see weather reports of their map area. *(Teacher Note: Tailor the specifics of this assignment to the level of understanding your students have of weather systems.)* If the state is in a different weather pattern zone from your own state, this could lead to discussions of weather science. Why is the weather different there? Are there mountain ranges between you and the map area that divert the flow of air? Etc.

Activity Six

Where Oh Where Does My Topo Map Belong? (Part 2)

This activity—like activity five—asks the students to research and investigate far beyond the scope of the topo map. In addition to learning some geography, students will see that their topo map relates to a real area with real people, cities, etc. Students will need access to other materials—almanacs, atlases, other maps, etc.—to do an effective job.

Figure 9 • Rain Shadow Effect
An example of how a landform—in this case a mountain range—affects weather patterns.

In this diagram, the prevailing wind blows from left to right, that is, west to east. Precipitation is usually heavy on the windward (in this case, the western) side of the mountain and light on the leeward side.

Using your large class map and other reference maps—such as those from an atlas, encyclopedia, or state highway maps—discover the general terrain of the state or region the topo maps represent. (Remember that terrain is the physical features of a section of land.) You can use longitude and latitude as marked on your topo map to accurately place the map (see activity four). Plains? Rolling hills? Ocean coastline—which ocean? Questions for students with more knowledge of botany and geography might include: What kind of plants are native to the area? What might families do for vacations there? What crops are grown? Are there economically valuable natural resources? How do people make a living there? Remind the students to use the many symbols on the map to assist them.

Perhaps the class could be divided into groups and each group given a certain topic to focus on in studying a topo map area. One group could study the political divisions (cities, counties, townships, etc.), another the topography and public lands (local, state, and national parks, etc.), and a third could study the economy of the area.

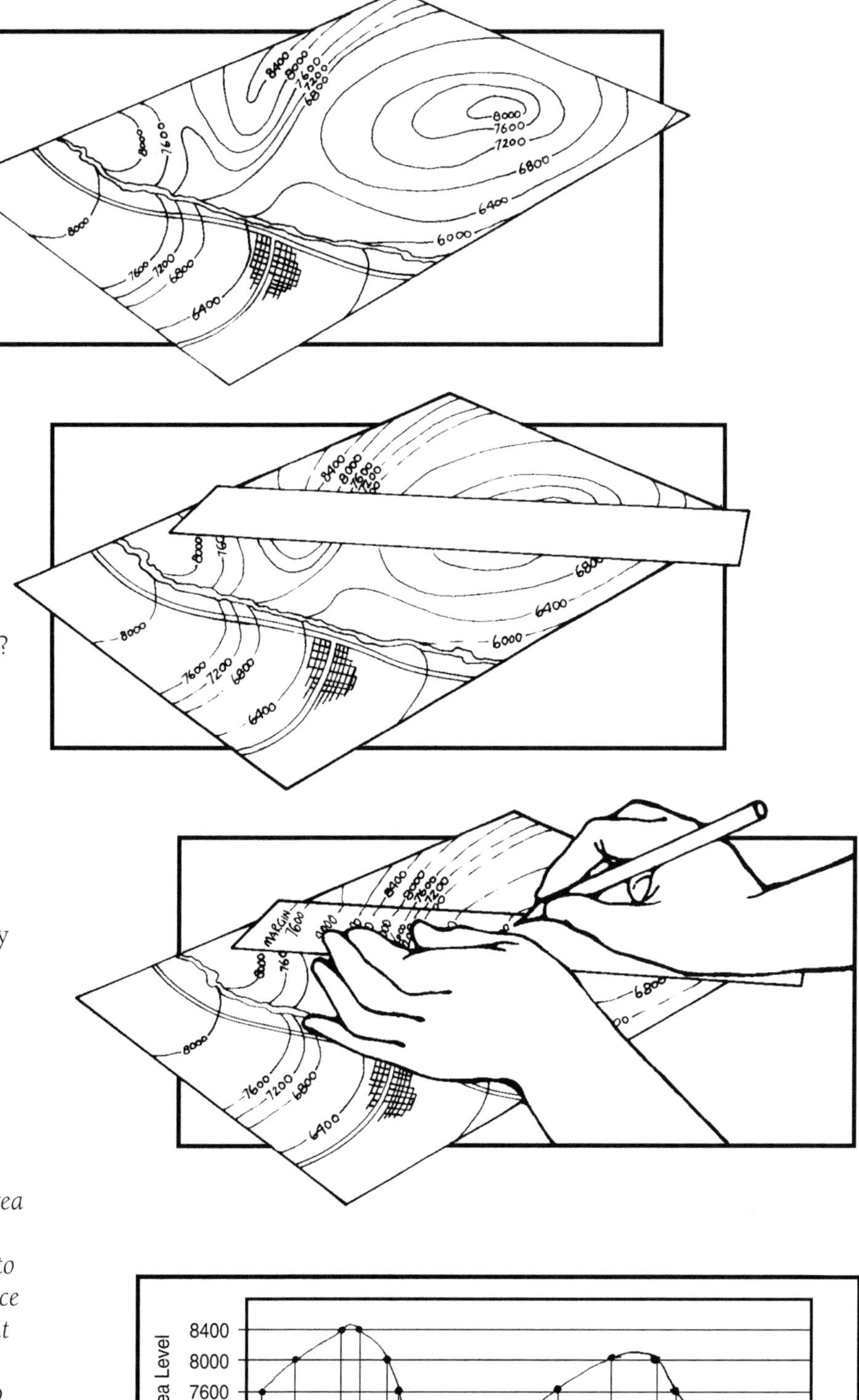

Activity Seven

Drawing a Profile

Students will draw a profile of a marked area on the map. This is similar to a geologic map's cross section. Students will need a 4 to 5 cm wide strip of blank paper, another piece of paper with straight lines running across it with the vertical scale marked in the appropriate elevations, and access to a topo map. Note: This activity results in a profile that is vertically exaggerated. It is meant to give students a sense of the lay of the land; it is not a rigid geologic cross section of the area.

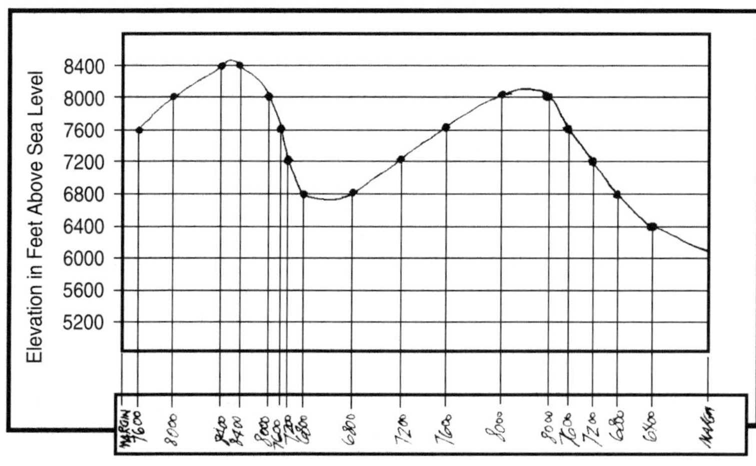

Figure 10 • Instructions for Drawing a Profile

Place the strip of paper across the topo map. Mark the strip of paper wherever the edge of the paper crosses a contour line. (You may want to use only index contours—the thicker contour lines—to simplify the exercise.) Everywhere there is a mark on the strip of paper, label the mark with its appropriate elevation. Move the strip of paper to the ruled paper and locate the points according to their elevation. The vertical scale will be determined by the range of elevations marked on the strip of paper. Connect the points by drawing a smooth line between them. (See Figure 10 on page 15.)

Activity Eight

Mapping Out a Play

This activity asks students to imagine and create. It may not be appropriate for all levels. Be careful that you give the students enough guidance about what you want as a result. Students will probably need access to a library or other resource books.

Work with your students to help them write a script for a play (one scene possibly) that would be acted out in the landscape of your topo map. The play should tell a story of some kind. Although the story doesn't have to be about the landscape, it is possible to work the landscape in as essential to the story. Help students visualize the landscape they would actually see if they were living at the map's location. Ask them how they can use certain features—mountains, rivers, buildings, wells, railroads—in the drama. Will you move from place to place in your drama? Will you have highway chases? Cowboy chases? If the map is near a large lake or along the coastline, is this a good map for showing places for a pirate to hide? It's possible the teacher and students could use the library to visually enhance the student's planning of this activity with books, films, atlases, etc. Examples of story lines for the plays include: an argument about who owns the abandoned mine (someone recently found gold in it), or a discussion by hikers/backpackers lost in the woods as to which direction they should go (see Figure 12 on page 20), or more involved plays about a pioneer family just arriving in the area or a Native American tribe that has been living in the area for centuries. If the topo map includes an urban area, the story could involve planning a new road in a congested area.

Activity Nine

Laying It Out

This activity asks students to analyze the "lay of the land" in the area the topo map represents. They will need to act as road planners and engineers to devise a road building strategy for the area. Some background material on road planning can be found in encyclopedias or other books. This activity could also lead into a discussion about the U.S. Interstate Highway System. The class will need enough topo maps for each group of two or three students (or quality copies) and possibly overlays (see discussion below).

Using your map, two or three students can plan the best way to lay out a new road through the landscape. The road can be a scenic route, or it can be a very practical way to link two places. Roads should be placed in the easiest, flattest places, and bridge construction should be planned if necessary. (Note: Road designers and planners know this too, so many of the flattest places on a map may already have roads.) It will be easiest to build a road up a slope if you go where contour lines are farthest apart. This indicates the most gentle slope. (Remember: the closer the lines the steeper the slope.) In addition, there will be no elevation gain if you follow the same contour line around a mountain. If your budget allows, each group of two or three students might design their road routes on plastic overlays which can then be compared with other overlays in the class. (You could use overhead projector film and tape together segments for your plastic overlay. Use overhead pens—with ink that wipes off—to trace the routes onto the overlays. This will allow you to reuse the overlay.) You could also use pieced-together photocopies of the original.

If your map has only smaller one-lane roads, try planning and building an interstate through the area. How wide is the right-of-way needed to build an interstate? Would you need to fill in any areas or build bridges? What buildings, agricultural, or undeveloped areas would be affected?

Students might also want to plan where houses, malls, dams, parks, or other areas might be built. They also might be given a role as a government official responsible for creating a National Park. What areas would they include? What facilities would the park need? If you have done or are doing any environmental studies, this might tie in.

Activity Ten

Down the Drainage Plan

This activity details the water and drainage system of the area your map represents. Some maps may be better than others at showing this, but most will have some streams or water on them. Three types of drainage systems are introduced. The activity also talks about the "rule of the V's." Read over the material below before doing this activity and make sure you understand the drainage patterns and rule of the "V"s. This activity is more of a demonstration and discussion than a hands-on activity. It also touches on meandering streams, a very useful topic to teach mathematical concepts in science. This exercise could supplement a geology section on river systems.

If your map shows a drainage system with rivers, lakes, or an ocean, have students plan a route people may have used to travel by water before roads were built. Ask the class who some of these people might have been? Students would start tracing at the higher elevations on the map and plan their canoe or boat trip to the lower elevations. In the process, they would trace out a drainage system from higher elevations and smaller tributaries to lower elevations and larger bodies of water.

Students will note that contour lines which cross rivers follow the "rule of the V's." The "V"s always point upstream (and uphill). We can even locate stream channels where no stream now flows, simply by finding the places where contour lines have these upstream pointing V's, one above another. Students can look for areas of V-shaped contour lines where the contour lines cross the stream, or mature valleys, where the flood plain is marked by widely spaced lines and is flanked by steeper elevations with closer contours at each side. Perhaps you can find evidence of a very broad, mature system, with abundant meanders, ox bow lakes, perhaps even swamps and yazoo tributaries. Additionally, can your students determine whether their drainage is dendritic (tree-like)? Radial (spokes on a wheel)? Trellis? (See Figures 11a, 11b, and 11c for examples of drainage patterns.) Are you fortunate enough to have any intermittent streams, disappearing streams, or springs?

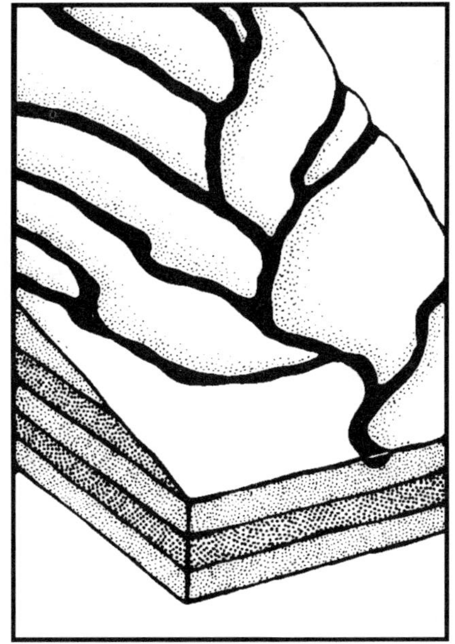

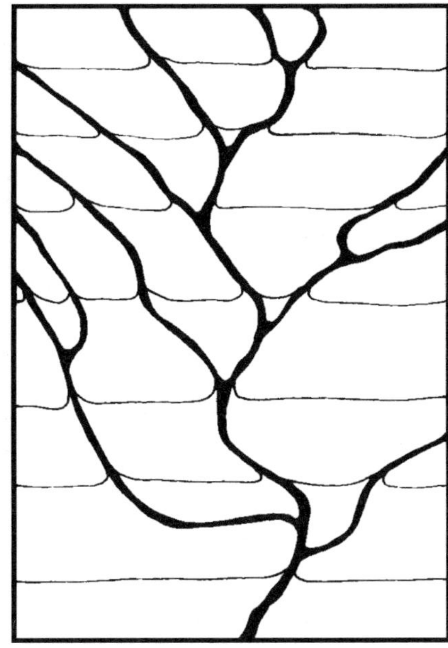

Figure 11a • Dendritic Drainage Pattern

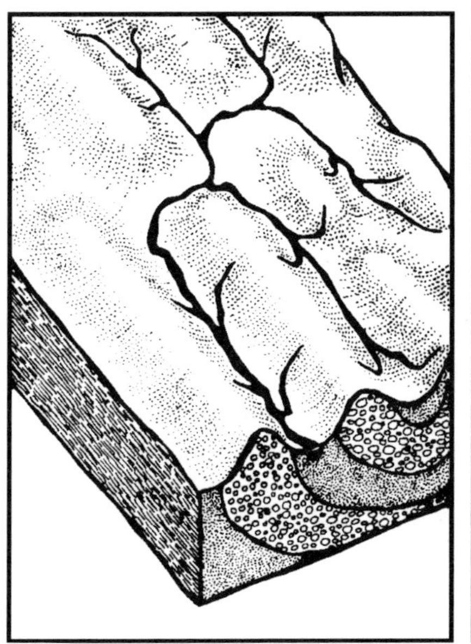

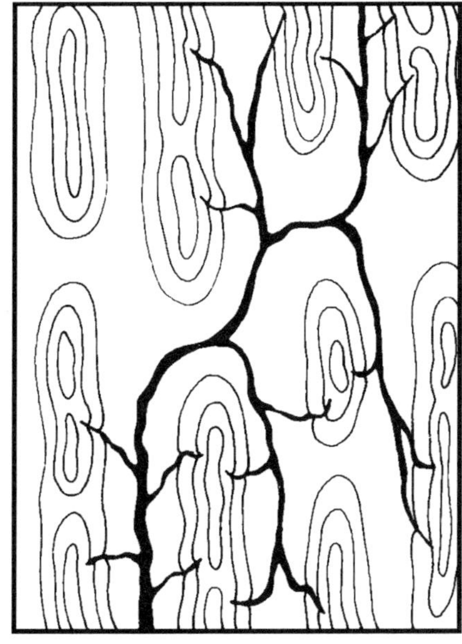

Figure 11b • Trellis Drainage Pattern

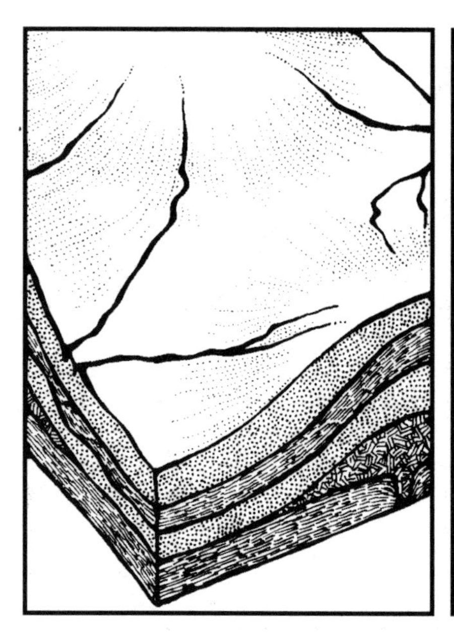

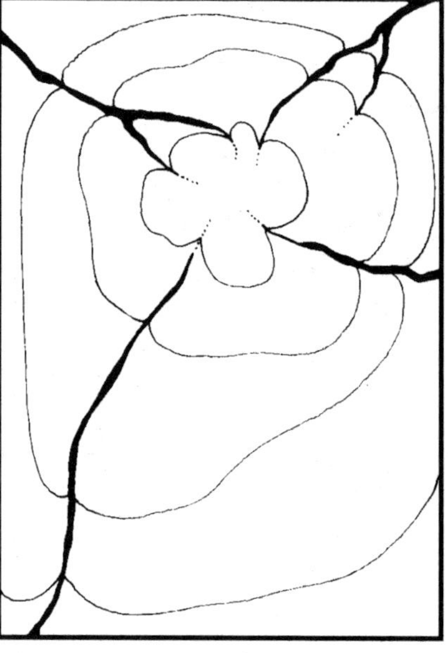

Figure 11c • Radial Drainage Pattern

Activity Eleven

Oceans to Mountains

This activity asks students to compare different areas of Earth and the way maps represent those areas. You will need to find other types of maps that show elevation or depth (such as bathymetric maps). (Even boating charts show depth measurements, usually called soundings.) Students should have access to these maps.

Perhaps your school has a relief map of a portion of the ocean floor showing features such as deep ocean trenches, the Mid-Atlantic Ridge, or the Hawaiian Islands. How could you imagine using the concept of contour lines to depict the ocean floor? There are at least two ways students might develop to do this. They might visualize using elevation lines like on a topo map to show height above the ocean floor. Conversely, they might think of measuring down, from sea level, to measure depth with contour lines. These examples could begin the imagining.

Look at a bathymetric map that shows contour lines of the ocean floor. The fathom is a common measurement, although newer maps now use meters. A fathom measures six feet. Ask students to compare this to a topo map. How are the two maps different? Ask students to think about how topo maps are now made. (It was discussed in the introduction.) Can these production methods be applied to charting oceans, bays, and other bodies of water? Would the charting be expensive?

Why don't nautical charts—which show water depth—look the same as topographic maps? Why don't they use contour lines to mark depths? (It would be quite a difficult task.) How would they go about mapping the areas? Can they use satellite mapping systems? What other technologies might they use?

Activity Twelve

Making Your Own Land

This activity asks students to create their own topo map. Younger students will need more guidance than older students. It is important to make sure students start out by deciding on a value for a contour interval. You may also want to set up some parameters as to what students need to include on their map, such as a lake, a mountain with one steep side and one gradual side, and maybe a town or roads.

Have students make up their own imaginary territory and make a map of it. Review some of the symbols students will have to use for their maps to be understood as topo maps. Contour lines close together will indicate steep slopes or mountains; widely spaced lines indicate gentle slopes. When students want to show valleys on mountains, the contour lines need to "V" upstream (see Activity Ten: Down the Drainage Plan) Students should be challenged to use as many of the USGS symbols as possible, to choose an appropriate contour interval, and to label the contour lines as they are on a USGS topo map. They may want to put in some benchmarks for elevation reference. Benchmarks are surveyors' reference elevation measures. A specific elevation, such as 3216 feet, is marked on a brass plate which is permanently set in the ground at that spot. They are useful points from which to measure. Have the students in one group swap their maps with the students in another. Have the second group try to describe the lay of the land.

Activity Thirteen

Mapping Out the Schoolyard

This activity involves outdoor schoolyard time. As with activity twelve, some basic parameters will be needed. Students will need measuring tools, such as a long tape measure (or even a long rope—which can be measured and used as a measuring device) and paper for note taking in the schoolyard. Check with the principal to see if there is a set of blueprints or plans for the school site that you could use to map out the schoolyard.

Map an area of your own schoolyard. This project will require choosing and measuring an area, and then relating that area to a piece of paper using a scale such as 1 cm to 100 cm. Choose an appropriate scale that will fit. Have students measure and place various objects on the map.

To determine elevation, students can use rise and run. Use a meter stick, a short ruler, and a small level. Attach the level to the meter stick with masking tape. Place one end of the meter stick at the highest point. Level the meter stick to the ground. Swing the meter stick around keeping it level, to determine what will be a logical contour interval. (A 10-cm drop will often work.) Place a marker where the short ruler touches the ground. Swing the meter stick around in a circle, marking points where it touches until there are enough points to create a contour line. Move the meter stick down to one of the other markers. Repeat the process for more contour lines.

A Classroom Discussion

A Topo Map Exploration

One of the authors, Dana Van Burgh, has been teaching topo map skills for many years. The following example from his advanced class combines some of the above activities into a integrated work period led by the teacher. This series of questions works well with students early in their work with maps. It forces them to closely examine the maps, and makes them use the map as a tool for further investigation. This series also works well as a review.

The following exercise can be done individually or in small groups. Teachers using this material will have to adapt it to the map or maps they own and to their area of the country. Please note that many of the answers to the questions that follow are "answers will vary." The answers *will* vary depending on the maps your class is using, grade level, and the students' background in maps. The exercise should result in an engaging discussion, not an oral quiz. Questions to the students appear in **bold typeface**, answers and comments—for teacher eyes only—appear in ***italicized typeface.***

What is the name of the map you have?

The name appears in two places, above the northeast corner and below the southeast corner. Remember, the top of the map is always north unless otherwise indicated. Look in the bottom margin of the map for a small diagram which is likely to have three arrows indicating the various norths. Note that all of them are toward the top of the map.

What does the "MN" stand for?

Magnetic north.

Why is it important for someone using this map to know the answer to that question?

So someone can compensate for the difference between magnetic north—the compass reading—and true north—what the map is based on. This compensation is called the declination.

When was this map published?

This will vary—look in the bottom margin, usually on the left side.

Why do we care?

Things may have changed since that time—new roads, towns, buildings, etc.

How many degrees north of the equator is the bottom of this map?

Remember, each degree contains 60 minutes and each minute of latitude (measurement north or south of the equator) is one nautical mile. Use the latitude markings at the corners to figure this out—it will vary for all maps.

Figure 12 • Hikers Using a Topographic Map

A little math. Using the information above, how many nautical miles from the bottom of the map to the equator?

Answers will vary.

A little more math. How many statute miles—the ones we use to measure highway distance—from the bottom of the map to the equator?

Answers will vary. (Teacher Note: Depending upon your class and the time and resources available, you may want to provide the two lengths. Most students should be able to find this without help.)

Just to let you check your work, two more math problems: How many nautical miles from the bottom of the map to the top?

Answers will vary depending on map scale.

How many nautical miles from the top of the map to the North Pole?

Answers will vary.

Reflect on the last few answers. Consider what you know about the angle between the equator and the pole and what you know about degrees and minutes and miles. Do your answers make sense? Do a proof.

Can you do this same process for longitude?

No, minutes of longitude vary in length from the equator to the pole.

Look at the map area. It is likely that there is a grid of squares drawn on it. Earlier you learned about Township, Range, and Section. These squares are the sections. (Remember, most areas east of Ohio probably don't have Township, Range, and Section coordinates.)

How big are the sections? If you don't remember, use the scale at the bottom of the map. (Look for the typical section. You may have noticed that some of them are a strange shape. Think about what might have caused this problem.).

Answers will vary depending on map scales and students' understanding of sections.

Now that you have a handy measuring tool, how wide is the area shown on this map? (That is the east-west distance.)

Answers will vary.

How far north-south?

Answers will vary.

Let's see if you can find the other pieces of the Township, Range, Section system. Look in either side margin, right along the edge of the mapped area. Someplace along here you will find a T and a number followed by either an N or an S. These are the township numbers. You should have found a couple of them close together and the line beside them is the boundary between townships. Use the same technique along the top or bottom to find the Range numbers.

Which section is in the southwest corner of your map? Remember, this is always written as section, then township, and then range.

Answers will vary, but the format should be Sec_____ T_______ R_______.

Now that you have two locator systems under control, it is time to consider the features on the map.

If there are green areas, what do they represent?

Vegetation, usually woodland. Check your key from the USGS for specifics.

Perhaps there are stippled areas. If so, what do they tell you?

Usually these areas represent sand or gravel. Again check the USGS key.

Everything in blue is water. Ponds are easy, as are streams, but what about streams that are not a continuous line? Some are indicated by short lines separated by three dots. What type streams are these?

These are known as intermittent streams—sometimes they flow, sometimes they are dry.

The most obvious printing on a topo map is the mass of fine brown lines running all over the place. What are these lines called?

Contour lines.

Why are some thicker than others?

The thicker contours are indexed—that is they should have an elevation number written along them somewhere.

What do the thicker lines have that the others do not?

Most have elevation labels.

Earlier you learned this simple rule for using topographic maps, "The closer together the contour lines, the steeper the slope." Use that knowledge to answer the following:

Are there many steep slopes in the area of your map?

Answers will vary.

Do they form any sort of a pattern?

Answers will vary.

Remember what you have learned about folding and faulting and erosion.
(Dana's classes learn about folding, erosion, and other geologic concepts before working with topo maps.)
Can you see any patterns of steep or gentle slopes which make you think about fault blocks?

Answers will vary.

Topographic maps are great for locating superimposed streams and water gaps. If you find any, give their location by Section, Township, Range.

Locations will vary, but they should in the proper format: Section______
Township_____
Range_____

Can you see anything about the hills and valleys or the streams, or other features, which might give you a clue as to why the population of the area is as large or as small as it is? Explain.

Answers will vary.

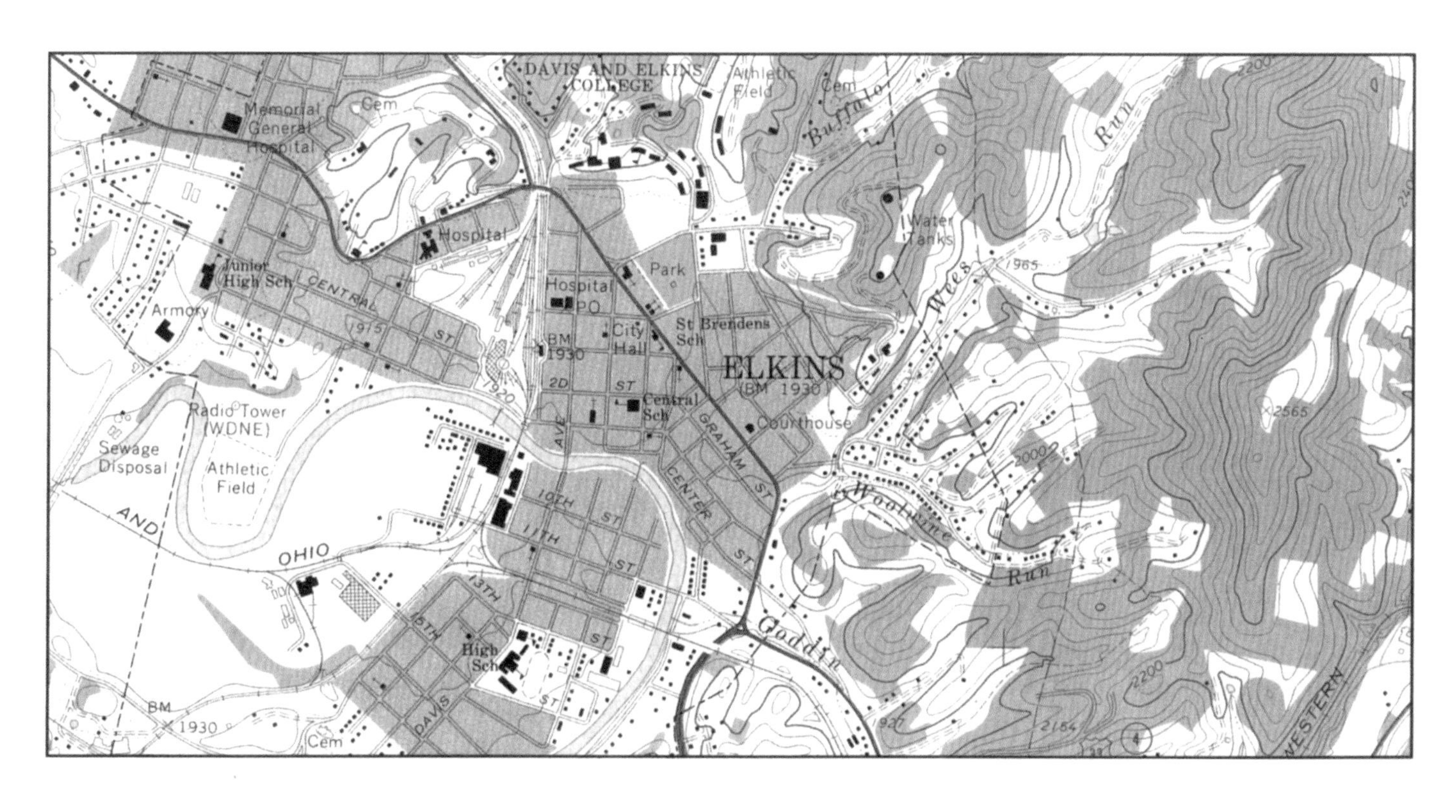

The Language of Topo Maps

Base line – In the land office grid system, a base line is a selected parallel of latitude used as a starting line for township—north/south— measurement.

Contour/Contour lines – lines on maps that pass through points of the same elevation.

Contour interval – the designated elevation difference between any two consecutive contour lines on a map. The interval can be different on different maps but it is always the same for an entire map. It is usually listed in the bottom margin at center of most topo maps, near the scales.

Declination – The angular difference between true north and magnetic north.

Dendritic drainage – A drainage pattern that looks similar to the branches of a tree—*dendro* means "like a tree"—as small streams and brooks converge into larger streams and possibly rivers. (See Figure 11a on page 17.)

Drainage pattern – the manner in which the streams, etc., in a particular area flow.

Elevation – the distance—in feet or meters—above sea level that a specific point lies. Contour lines indicate elevation.

Index contour – The thicker brown lines on a topo map. These lines usually have a number along them that denotes the elevation along the line.

Intermediate contour – the thinner brown lines between index contours.

Latitude – A measurement of arc on Earth. Measured in degrees north or south of the equator, lines of latitude are called parallels.

Longitude – A measurement of arc on Earth. Measured in degrees east or west of the Prime Meridian, lines of longitude are called meridians.

Nautical mile – Equal to one minute of latitude—6080 feet (1853.2 meters).

Oblique – From an angle that is neither parallel or perpendicular.

Prime meridian – The line of longitude which runs from pole to pole through the original site of England's Royal Observatory in Greenwich. All other meridians—lines of longitude—are measured east of west of the Prime Meridian.

Principal meridian – In the land office grid system, a principal meridian is a selected line of longitude used as a starting line for range—east/west—measurement.

Quadrangles ("quads") – Term for a topographic map with four equal sides. For example, a 7.5 minute-series map is a quadrangle, since it measures 7.5 minutes of latitude by 7.5 minutes of longitude.

Radial drainage – A drainage pattern characterized by streams that run away from the center of a high circular or oval landform, such as a dome or a volcano. (See Figure 11c on page 18.)

Range – In the land office grid system, range is a six-mile unit of measure east or west of a principal meridian.

Rule of the V's – Phrase used in topographic map study that refers to the characteristic "V" shape a contour line has when it passes through a stream. The bottom of the "V" always points *up*stream.

Scale – Relationship between the distance on a map and the distance it represents on Earth. For example, a scale of 1:24,000 means one unit on the map equals 24,000 of those units when measured on Earth.

Section – In the land office grid system a section is a one-mile by one-mile area of land. There are 36 sections in a township.

Topographic map *(also known as topo map)* – A map that uses contour lines and symbols to represent the features—man-made and natural—of the mapped area.

Township – In the land office grid system, township has two meanings. Township is measured north or south of a selected base line. **A** township refers to an actual six-mile by six-mile plot of land.

Trellis drainage – A drainage pattern characterized by streams that drain parallel valleys. (See Figure 11b on page 17.)

USGS – The United States Geological Survey is the branch of the federal government's Department of the Interior responsible for creating many types of maps, including topo maps.